KNOWLEDGE ENCYCLOPEDIA

HUMAN BODY
BRAIN & NERVOUS SYSTEM

© Wonder House Books 2024

All rights reserved. No part of this book may be reproduced or transmitted in any form by any means, electronic or mechanical, including photocopying and recording, or by any information storage and retrieval system except as may be expressly permitted in writing by the publisher.

(An imprint of Prakash Books)

contact@wonderhousebooks.com

Disclaimer: The information contained in this encyclopedia has been collated with inputs from subject experts. All information contained herein is true to the best of the Publisher's knowledge.

ISBN : 9789389931259

Table of Contents

Making Sense of Who We Are	3
Command and Control: The Brain, Spine and Nerves	4
Action and Reaction: Sense and Motor Organs	5
Fitting Together: Moving Pictures Sent on Radio Waves	6–7
Putting Matters in Grey and White	8–9
Voluntary and Involuntary Movements	10
Moves Like Lightning: The Autonomic Nervous System	11
Breaking News to the Brain: The Senses I	12–13
Sight and Sound: The Senses II	14–15
Nerves and Neurones: How They Work	16–17
Your Body's Other Brain	18–19
Switching Your Nerves Up and Down	20–21
From Thought to Action	22
Making Memories	23
Higher Thought: From Animal to Human	24
The Pituitary and Pineal Glands	25
Higher Thought II	26–27
An Attack of the Nerves: Epilepsy & Multiple Sclerosis	28
Sleep Healthy and Wake Up Healthy	29
A Healthy Mind in a Healthy Body	30
Improve Your Mental Health	31
Word Check	32

MAKING SENSE OF WHO WE ARE

Our brain is a complex organ. Scientists are still understanding how it works. We know that the brain takes information from the eyes, ears, tongue, nose, and skin to the spinal cord. It sends this information back to the brain through the sensory nerves. Through many steps, the brain puts together the information from the senses and makes up the world that we live in—our thoughts, memories, likes, and dislikes.

Based on what it sees, hears, tastes, smells, touches, and feels, the brain passes information through the motor nerves to the muscles and organs. Which is why we feel hungry, frightened, angry, thirsty, etc. For instance, when we are faced with a difficult situation, our brain tells us whether we should run away or stay and fight.

The nervous system also controls what we do without being aware of it, like breathing, circulating blood, digesting food, and removing waste from the kidney. Doctors call this part of the nervous system the autonomic nervous system.

▶ *Our brain remakes the universe inside our heads*

Command and Control
The Brain, Spine, and Nerves

The nervous system is to our body what the government is to a country. The nervous system takes the information coming from various sensory organs, processes it and makes decisions, which are then carried out by other organs. The nerves, which are a part of the nervous system, are made up of special cells called **neurones** that take tiny electrical signals from the sense organs to the brain and from the brain to the motor organs. Then the motor organs, mainly the muscles, act together to do what the brain wants them to. The neurones communicate through tiny switches called **synapses**, and chemicals that doctors call **neurotransmitters**.

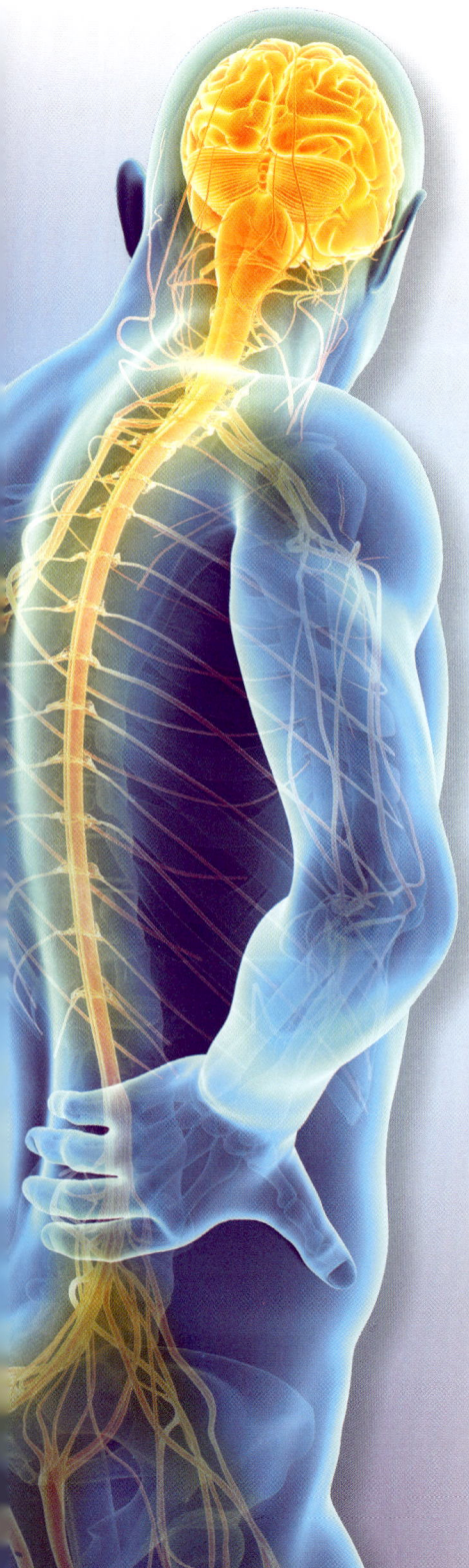

The Brain

The human brain is made of **white matter** and **grey matter**. As human beings evolved, they grew bigger brains with more neurones. To fit them all in, the grey matter had to fold itself into wrinkles. The neurones have connections with each other that run all over the brain. This makes up white matter. Inside the spinal cord is the central canal which contains the cerebrospinal fluid. This supplies the brain and spinal cord with nourishment.

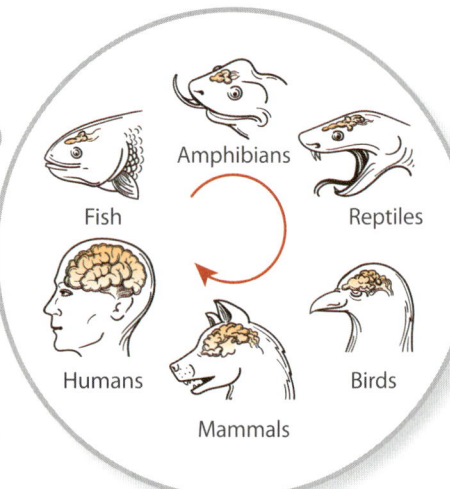

▲ *The grey matter does the processing, whereas the white matter provides communication*

The Spinal Cord

The spinal cord starts at the back of the neck and extends to below the last rib. It takes information to the brain, and brings back orders from the brain to the body. It also takes care of the body's reflexes. The spinal cord is protected by the backbone which starts at the base of the skull and ends above the hip area.

The nerves work like telephone cables that carry messages from the organs to the brain and back. The organs in the head are directly connected to the brain, while those in the body connect to the spinal cord.

◀ *The brain and spinal cord govern the rest of the body. Nerves connect them all together*

In Real Life

Look closely at the pictures of a walnut and brain. They look a little alike because like the brain, the walnut has folds and is divided into two halves. Additionally, eating walnuts is good for the brain. Walnuts are rich in minerals and lipids that help your nerves and brain function well.

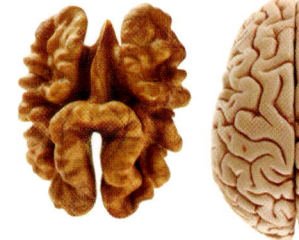

▶ *Some scientists claim that walnuts are the healthiest nuts of them all*

Action and Reaction
Sense and Motor Organs

Imagine a pineapple cake in front of you. Your eyes give your brain a picture—it is big, round, and white. Your nose tells your brain that it smells of pineapples and cream. These are called stimuli. Your brain puts all the stimuli together and makes your hands reach out to grab a piece of the cake! As your fingers touch it, your brain knows that it is creamy on the outside and crumbly on the inside. You put it into your mouth, and your tongue tells you that it is sweet to taste. How did it all happen?

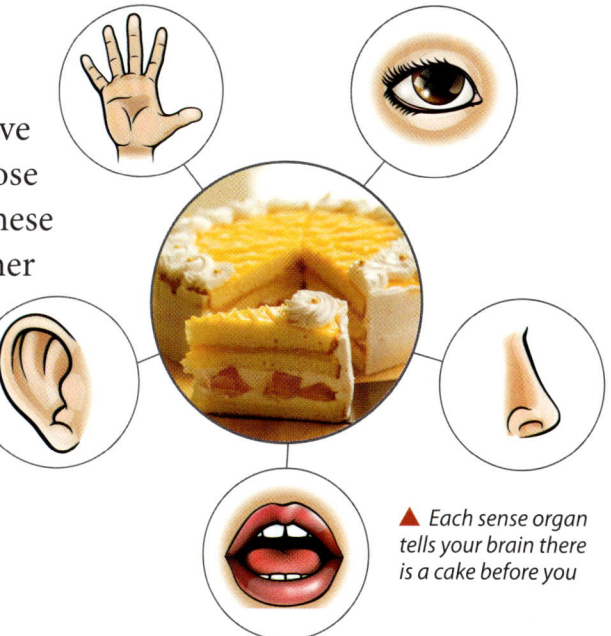

▲ Each sense organ tells your brain there is a cake before you

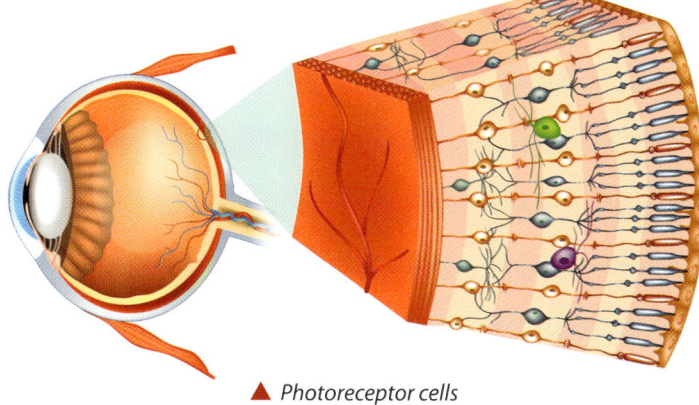

▲ Photoreceptor cells

Sense Organs

Each of our senses works in their own way. Chemoreceptors in the tongue and nose can sniff out the smallest amounts of chemicals in our food and air. Photoreceptors in the eyes pick up colours and light. Auditory receptors in the ears pick up the tiniest vibrations in the air. All of these turn into tiny electric signals that go through the nerves to the brain, which turns them into tastes and smells, sights and sounds.

Motor Organs

When your sense organs tell your brain that there is a cake in front of you, your motor organs take action to grab a slice of the cake. Your brain sends signals through your nerves to your hand muscles, so they reach out for the cake. It also sends signals to your mouth and your voice box, so you can ask the grown-up in the room for permission to grab a slice!

💡 Isn't It Amazing!

We cannot sense electricity by looking at a wire. But bees can detect the tiniest of electric currents from the flowers they visit. A bee can even sense whether another bee has already been to a flower that it is visiting. What's more, it comes to know whether the first bee took nectar from the flower. This saves the worker bee's time!

▲ Bumblebees can sense electricity

▲ Your motor organs act on the information they receive from the brain and the sense organs

Fitting Together
Moving Pictures Sent on Radio Waves

The brain is divided into many parts. These include brain stem, little brain, midbrain, and higher brain. They work like a jigsaw puzzle where all the pieces make sense only if they fit together.

▼ *The diagram shows the parts of the inner brain*

The brain is enclosed within the skull, which protects it from injury. It is also covered by three membranes called **meninges**. The meninges help keep the brain cushioned, so even if you hurt your head, the brain does not smash against the inside of your skull. Veins and arteries criss-cross the meninges, bringing nutrition to the brain, and taking away any waste.

👤 The Little Brain

This part of the brain is just above the back of your neck and makes up about one-tenth of the brain. Doctors call it the cerebellum. It acts like a checking centre for the brain, helping muscles correct themselves. For example, if something seems heavier than it looks, your cerebellum will get your muscles to put in more force. The little brain also helps you learn to avoid things that cause pain.

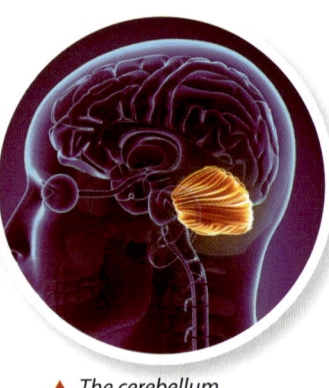

▲ *The cerebellum contains significantly more nerve cells than the cerebrum*

👤 The Brain Stem

The brain stem is the first part of the brain. It is made of two parts—the **medulla oblongata** connects the spinal cord and the brain, while the pons connect the cerebellum and the cerebrum to the medulla. Between them, they control our daily activities like the beating of the heart, the circulation of blood, and the working of the lungs.

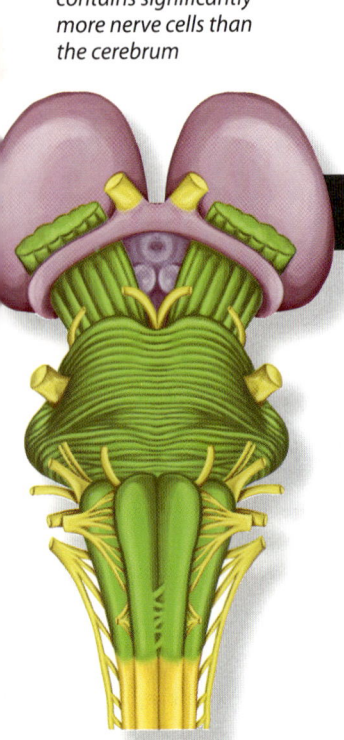

◀ *Structure of the human brain stem*

The Higher Brain

The rest of the brain makes up the cerebrum. Most of what you see is the outer **cortex**, divided into two brain hemispheres (right and left) by the longitudinal fissure. The cortex is made up of four lobes, divided by inward folds called **sulci**. The outward folds of tissue that make up each lobe are called **gyri**. Each lobe has a function and also helps the other lobes.

The temporal lobe is responsible for hearing, memory, and learning. The occipital lobe is responsible for the function of sight. The parietal lobe is responsible for touch and movement. The frontal lobe is for thinking, acting, language, and personality.

In Real Life

A few decades ago, doctors would prescribe cutting off bits of the brain to cure moral deviancy and brain diseases like epilepsy. They called this lobotomy. More often than not, it left the patients worse than before. Often, after a lobotomy, a patient was unable to show emotion. Modern doctors think it was a cruel and useless practice.

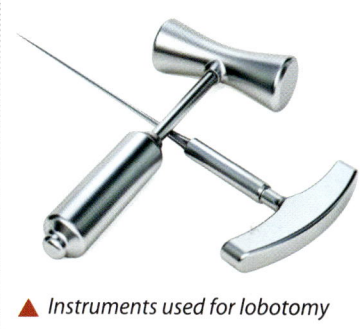
▲ Instruments used for lobotomy

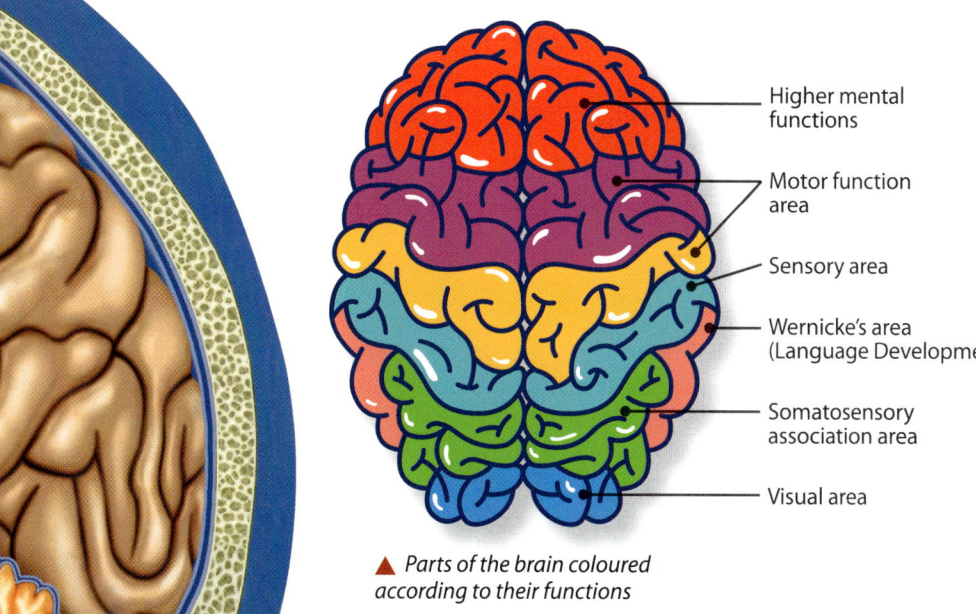
▲ Parts of the brain coloured according to their functions

Cerebellum

The Inner Brain

The inner brain is made up of the **thalamus**, the **hypothalamus**, and the midbrain. The thalamus is the part that decides which stimuli to give attention to. This is the one that says, 'If you see a tree, do nothing. But if you see a lion, you better run!' The hypothalamus takes care of things like feeling hungry or thirsty, tired or fresh, sleepy or awake. The midbrain connects the senses, if you hear mum's voice, you immediately turn to see where she is.

★ Incredible Individuals

The procedure of lobotomy was developed by Antonio Egas Moniz but it was American neurologist Walter J. Freeman II, who popularised its use despite widespread criticism. He approached the media to promote the procedure, even giving it the name 'prefrontal lobotomy'. During his time, he performed around 3,500 lobotomies. Thankfully, he was banned from doing it after the last lobotomy, which happened in 1967

▲ Antonio Egas Moniz

▲ Walter J. Freeman II

Putting Matters in Grey and White

Ever heard of the phrase, 'use your grey matter'? That is the part of the brain responsible for thinking, remembering, and imagining. The brain is made of billions of tiny cells called neurones, that act like electric cables. Each neurone has a cell body, and long, wire-like dendrites and axons. The cell bodies of neurones clump together into tissues called grey matter.

The axons of neurones, tied into long bundles called tracts, make up white matter. They connect the different parts of the brain so that the grey matter can do all of its work.

Grey Matter

In the cortex, grey matter is present as gyri. In the inner brain, it is made of clumps known as **basal ganglia** or **nuclei**, surrounded by white matter. Each of these has a name and a specific function to perform. These nuclei let your brain decide how to act once you have a stimulus, like which way to turn when you want to catch a ball. They are also important for learning and memory and to control emotions and behaviour.

The most important is perhaps the claustrum. Famous scientist Francis Crick compared it to the conductor of an orchestra as it connects with all parts of the brain.

◀ Grey matter is made of cell bodies of neurones while the thread-like white matter is made of their axons

Nucleus	What It Controls
Caudate nucleus	Planning movements, memory storage, learning, and decision-making
Putamen	Voluntary and **involuntary movements**
Globus pallidus	Voluntary movements
Subthalamic nucleus	Voluntary and involuntary movements, learning, emotional reactions
Substantia nigra	Eye movements, mood, learning, likes, and dislikes
Claustrum	Consciousness

White Matter

The biggest tract of white matter joins the right and left halves of the cortex. Doctors call it the corpus callosum. This lets the nerves in each half talk to each other. Other tracts connect the basal ganglia with the cortex, the midbrain with the basal ganglia, and the hindbrain with the midbrain. The pons is also made of white matter. One important tract connects the eyes with the occipital cortex, crossing over in the middle, so that the left eye connects to the right side and the right eye to the left side!

In Real Life

Atrophy means the shrinking of a tissue. People who are addicted to alcoholic drinks get affected by atrophy of the white matter in their brains. This makes them forget things and they are unable to move their arms and legs properly.

▲ Alcoholism doubles the risk of brain shrinkage in the 30s to 50s age groups

Brain Ventricles

Like the heart, the brain also has four '**ventricles**' filled with cerebrospinal fluid that keep it nourished. There are two lateral ventricles on the sides, and the third ventricle and fourth ventricles connect them to the central canal of the spinal cord. A thick network of blood capillaries in the choroid plexus supplies the cerebrospinal fluid (CSF) with the minerals and nutrients that it needs.

Ventricle from side view

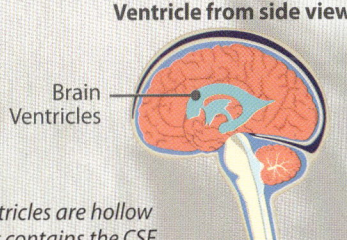

Ventricle from top view

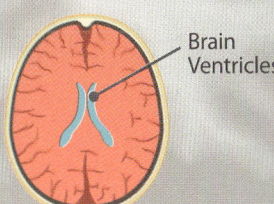

▶ The ventricles are hollow spaces that contains the CSF

Incredible Individuals

Many mental and nervous diseases like Tourette (pronounced as too-ret) syndrome happen because of damage to the basal ganglia. A patient of this disease appears quite normal, but may repeat themselves while speaking or make sudden sounds while talking, including grunts, whistles, or swears, or make sudden movements called tics. A lot of what we know today about this syndrome comes from Oliver Sacks (1933–2015), a British neurologist who researched on unusual and rare neurological disorders. He received the Guggenheim fellowship for his research on Tourette syndrome.

▶ Oliver Sacks' research helped many patients with this syndrome. David Beckham and Wolfgang Mozart have suffered from Tourette syndrome

Voluntary and Involuntary Movements

Did you know that our body is in motion all the time? But we are aware of only some movements, like walking and eating. We can control these movements when we like. When you are lagging in a race, you can make your leg muscles work harder so you run faster.

▲ Voluntary movements done without thinking are called reflexes

Understanding Body Movements

Some movements in the body happen on their own and we do not have control over them. For example, our heart beats and our stomach churns. These are called involuntary movements. Your heart beats faster when you are excited or frightened, but you cannot make it beat slower even if you want to. Conversely, some voluntary movements can become involuntary when you have to react very fast. For example, if you burst a balloon behind someone, they cover their ears. Some involuntary movements can become voluntary. Breathing, for example, is an involuntary action. Your brain regulates your breathing; specifically, the part of your brain called medulla oblongata. However, if you run too quickly, you might need to take in greater gulps of air to feel better. At this point, your breathing becomes voluntary. This is controlled by another part of the brain called the precentral gyrus.

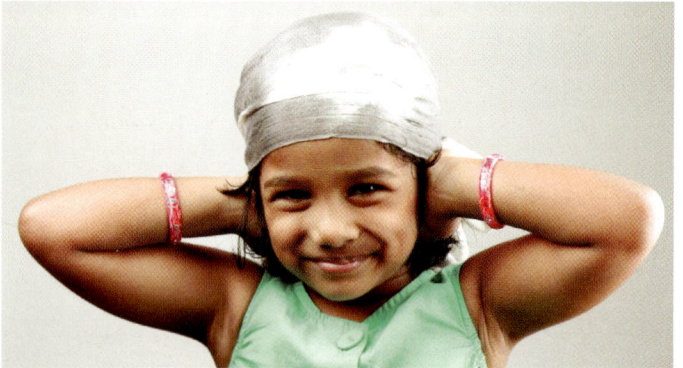

▲ When you hear loud or sudden noises, your hands might reflexively cover your ears

Isn't It Amazing!

Gurning is the British slang word for making faces. If you can make a lot of faces, you should head to the Gurning World Championships in Egremont, UK. The top prize goes to men, women, boys, and girls who can make the most faces.

▲ Voluntary movements are controlled by the higher brain

Voluntary Movements

These are movements that are controlled by the higher brain, especially the sensorimotor cortex in the parietal lobe. These include control of the muscles of the legs and arms, fingers and toes, face and jaws, and the abdomen. That is why you can smile at friends or draw your stomach in when you have to stand in attention.

Involuntary Movements

These are the movements controlled by the cerebellum and lower nuclei of the brain and also by the autonomic nervous system. These include the contraction of the stomach and intestines to move food forward, the beating of the heart, the blinking of the eyes, and the contraction and expansion of the lungs.

▶ While closing her eyes and sniffing, this girl has made her involuntary muscles become voluntary

Moves Like Lightning
The Autonomic Nervous System

Sometimes you need to react so fast that there is not enough time for your brain to think. That is when the autonomic nervous system takes over. For example, when you accidentally touch a hot pan, your hand instantly withdraws away from it; or when ambient light shifts, the dilation of the pupils will change accordingly.

It also takes care of the things which would otherwise take too much of your time, such as digesting food, making blood flow through your body, helping your tired muscles become relaxed, and keeping your urinary bladder in check when you are sleeping. When you wake up, it lets the brain take over. The automatic nervous system is divided into two parts: the sympathetic and parasympathetic nervous systems.

In Real Life

Sympathy has nothing to do with the sympathetic nervous system! Instead, it is controlled by a part of the brain's temporal lobe called the **amygdala** and special kinds of neurones called **mirror neurones**.

▲ Did you know that children can feel sympathy from the age of two?

 ## Fight or Flight

The spinal cord and a set of nerves, which constitute the sympathetic nervous system take care of most situations when you need to act without thinking too much. It prepares your body to either run away from a dangerous situation, or to turn around and fight. So, if a mosquito sits on your hand, you might swat at it without thinking. In more serious situations such as people getting into fights or discovering a fire, they might punch the opponent or douse the fire. When this system is active, you feel emotions like excitement, anger, fear, and disgust.

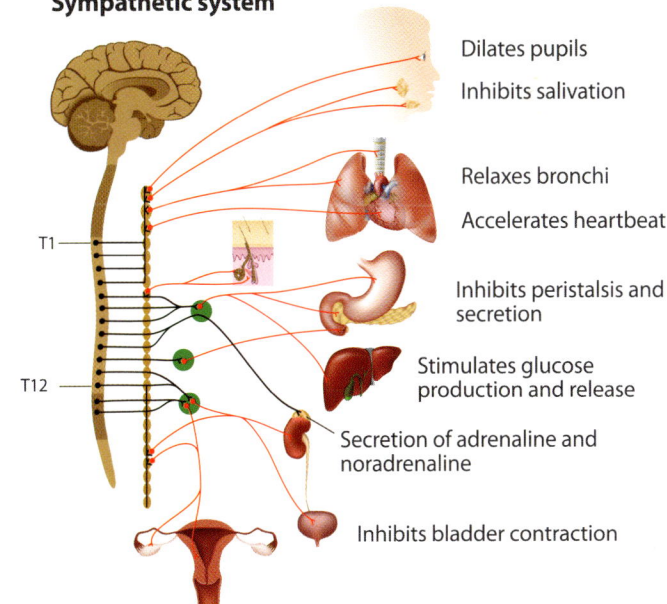

▲ The parasympathetic nervous system prepares your body for rest and digestion

▲ The sympathetic nervous system prepares your body to either fight something or run away from it

 ## Rest and Digest

The parasympathetic nervous system makes you feel tempted to eat when you see the food you like. It makes you calmer and happier, while your digestive system gets ready to digest. You start drooling, your stomach and small intestine start releasing digestive enzymes, and your large intestine prepares to get rid of your last meal so that there is space for the new meal. When this system is active, you feel satisfied and sleepy.

Breaking News to the Brain
The Senses I

All our senses can be divided into two. There are senses that need to be felt, such as touch, taste, and smell. Then, there are senses like sight and sound that are not felt. You can also divide them into those that help you ascertain the direction of the sensory stimulus (as shown on this page) and those that do not (see *pp 14–15*). The sense organs in the head communicate directly to the brain, while those in the rest of the body (mostly in your skin) communicate with the spinal cord. The sense of balance is unique as it does not depend on external stimuli.

 ### Taste

If you look at your tongue in the mirror, you will see that it is made up of hundreds of grainy bumps that doctors call **papillae**. Each papilla has a number of special nerves called **gustatory receptor cells** (GRCs). Different papillae have different kinds of GRCs for things that taste sweet, sour, bitter, or salty.

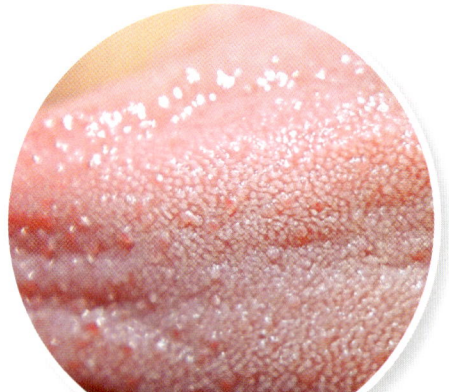

◀ The tongue has around 3,000 to 10,000 taste buds

 ### Smell

Deep inside the nose is an organ called the **olfactory bulb**, which gives us the sense of smell. It is made of thousands of **olfactory receptor cells** (ORCs), which catch chemicals in the air we breathe and tell us whether they are nice, like oranges, or pungent, like garlic. The nose also has pain receptors, which help pick up very strong smells that suggest that you may be exposed to a dangerous chemical like ammonia.

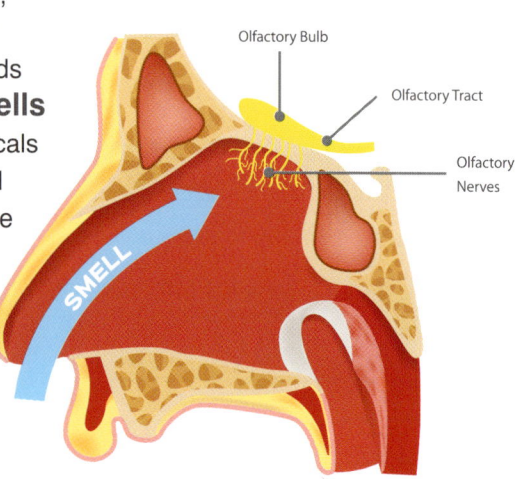

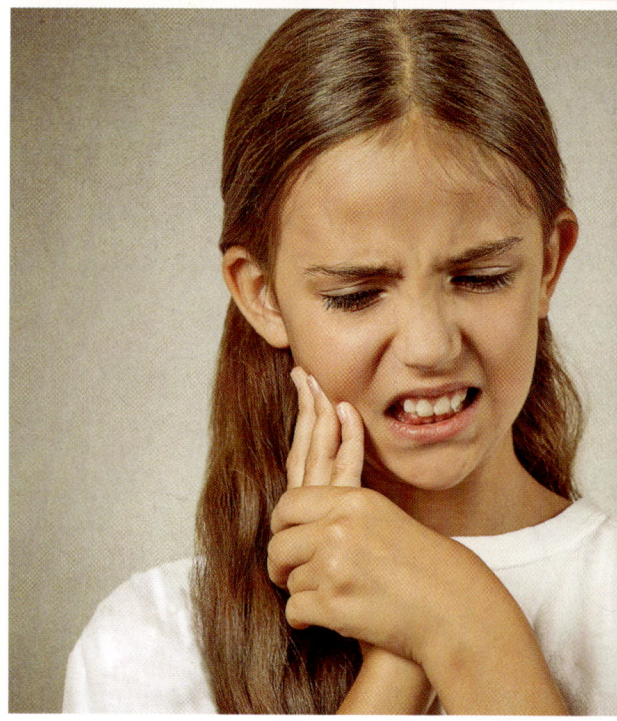

HUMAN BODY — BRAIN & NERVOUS SYSTEM

Isn't It Amazing!

Butterflies have their taste buds on their feet!

▲ Butterflies put their feet on the flower to quickly decide if it has nectar

In Real Life

A lot of what we think of as 'flavour', is actually both smell and taste together. The brain treats them together. That is why, when you have a cold and your ORCs are blocked due to a runny nose, food tastes odd even though your GRCs are not blocked.

▶ Having a cold prevents you from tasting or smelling things properly

Touch, Heat, and Pain

Our skin is a giant sensory organ that can feel many things. Different kinds of nerve endings in the skin tell the brain and spinal cord different things. Each part of the skin is represented in the brain in a map called the homunculus.

▶ Skin is the largest organ in the body

Cutaneous receptors

Sensory Ending	Place in the Body	What it Senses
Free nerve endings	All over the body	Pain, heat/cold, pressure, or twisting
Merkel's discs	All over the body	Very light touch (like a mosquito landing)
Ruffini's corpuscle	Skin and joints	Stretching the body
Meissner's corpuscle	Fingertips and lips	Light touch
Pacinian corpuscle	Inner skin	Strong pressure (like a tight hug)
Root hair follicle plexus	Hair follicles in the skin	Movement of hair (like an insect brushing past it)
Krause's end bulbs	Eyes, lips, and tongue	Cold

Incredible Individuals

Anaesthesia is a medical treatment given to a patient before surgery by doctors, so that one does not feel any pain at all. It was invented in England in the 19th century, but many doctors did not use it, making their patients go through surgery in great pain. Queen Victoria (1819–1901) decided to take anaesthesia when she gave birth to her eighth child, Prince Leopold (1853–1884). This made the public demand it and doctors began to use it more.

▶ Queen Victoria was the first member of the Royal family to live at Buckingham Palace

Sight and Sound
The Senses II

Our eyes and ears do not just see and hear. Since we have two of each, the brain can also ascertain where the stimuli are coming from, using the eyes and ears. So, if you hear something falling nearby, you know that you have to run in the opposite direction to save yourself. When you are playing cricket, your eyes tell you how far away the ball is from where you are standing, so you know when to hit it. Your sense of balance allows you to walk smoothly and correct yourself if you slip.

Hearing

The ear 'hears' just like a telephone does. The outer ear acts like a microphone, collecting sound from around you. These sounds go through your ear canal to the ear drum. This is a tiny stretch of skin that vibrates, just like a telephone diaphragm. The tiny ear bones pick these vibrations and take them to the cochlea. The cochlea is coiled like a snail's shell and breaks what you are hearing into separate sounds, so you can hear even if separate people are calling for you at the same time.

The brain can hear the same sound through both ears, so it can figure out whether the sound is coming from the left, the right, the front, or behind you.

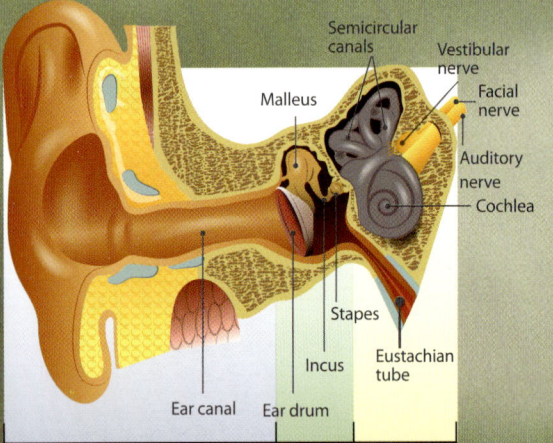

▲ Your outer ear never stops growing throughout your lifetime

Balance

Do you see the semi-circular canals in the figure above? These have nothing to do with hearing, but tell your brain about your body's balance. They are filled with a jelly-like substance and have tiny hairs inside them. If you are wobbling without balance, the jelly moves and the hairs pick up on this movement and communicate to the brain. The brain then sends a message to the relevant muscles to steady yourself, even without you realising it.

Incredible Individuals

The famous music composer Wolfgang Mozart had a rare ability called an 'absolute pitch', which helped him identify correctly a musical note and even compose the same himself.

▲ Mozart composed his first piece of music at the age of five!

Sight

The human eye works just like a camera. The pupil and iris in the front of your eye act like shutters that close and open to let in light. The lens concentrates light just like the lens of a camera. Light finally lands on the retina, which has special **photoreceptor cells** (PRCs). They are of two types—**rod cells**, used for night vision; and **cone cells** that work better in daylight. The cone cells sense only three colours of light: red, green, and blue.

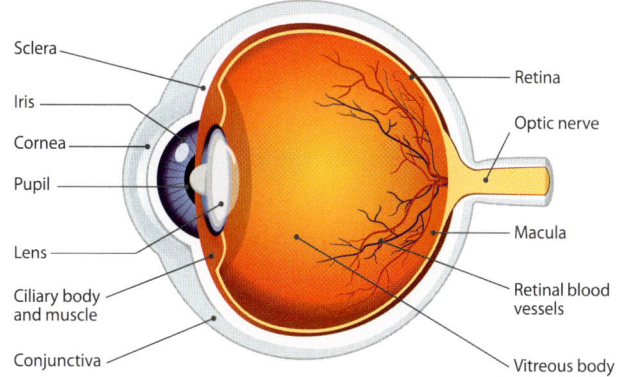

Binocular Vision

Both your eyes tell the brain slightly different things. What is right in front of you is the same, but the left eye sees things to your left, and the right eye sees things to your right. Using these differences, the brain can make a 3D map of the world in front of you, so not only do you see things, but you also know how far they are.

▶ Humans have a maximum horizontal field of 200° with both eyes

In Real Life

Can you spot the numbers in the picture? Colour blind people cannot as they have defects in their cone cells that stop them from sensing the difference in colours.

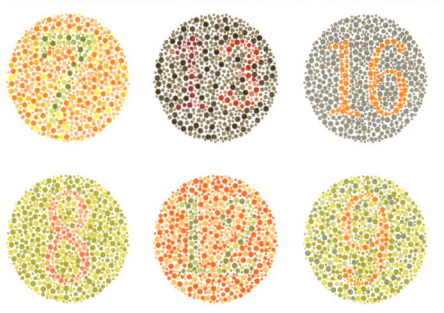

▶ The numbers are written in different colours than the circles

Nerves and Neurones
How They Work

The neurone is a nerve cell that forms the basic unit of the nervous system. Unlike other cells, neurones can stretch for several inches along the body. They clump together in the brain to make nuclei. In the rest of the body, neurones clump together to make **ganglia**. Bundles of neurones that interact with the same organ are called nerves.

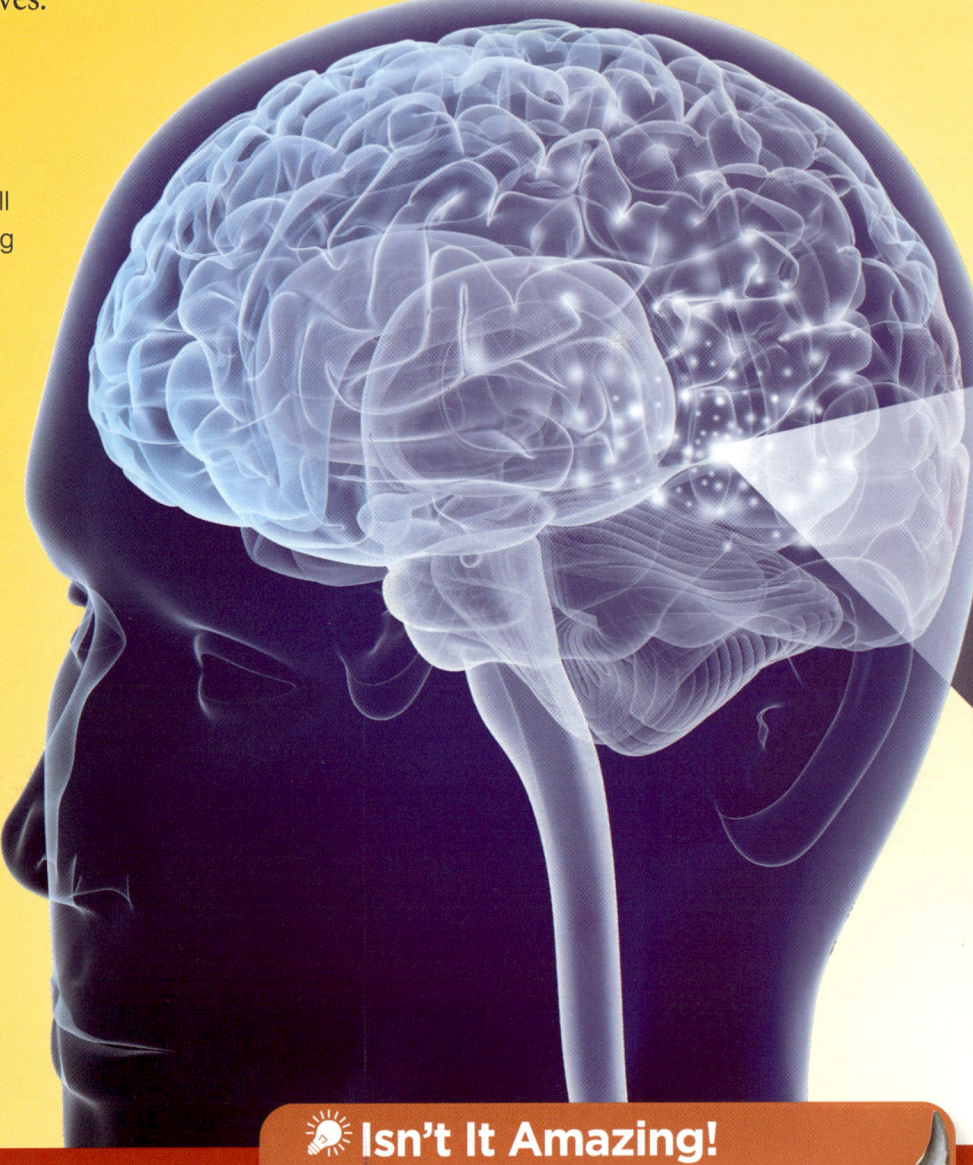

▼ The diagram provides a closer look at the nerves, neurones and axons of the body

👤 Cell Body

Most of the neurone's mass sits in the cell body. This has lots of mitochondria, giving the neurone all the energy it requires.

👤 Dendrites

Dendrites look like plant roots and are parts of the neurone that touch other body cells or neurones. They gather information from the sensory cells and pass them onto the axon at the other end.

👤 Axon

This is the longest and most important part of a neurone. It carries news from the dendrites to the next neurone over long distances. This information is carried in the form of tiny amounts of electricity called an action potential. **Axons** in the brain make up its white matter and are not covered by myelin sheaths.

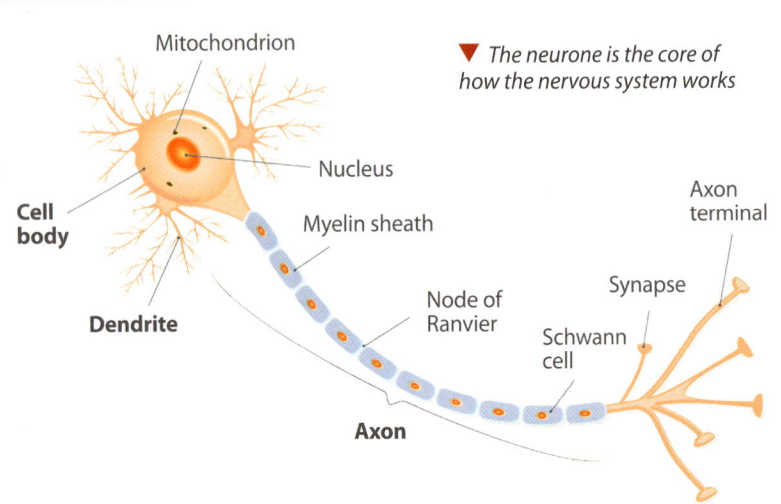

▼ The neurone is the core of how the nervous system works

💡 Isn't It Amazing!

A blue whale can grow up to 34 metres long. Some of the neurones in its spinal cord run all the way to the tip of its tail. At nearly 30 metres, this makes the blue whale's spinal neurones the longest single cells in the world!

▲ Blue whales have the longest neurones in the world

HUMAN BODY — BRAIN & NERVOUS SYSTEM

Myelin Sheath

Like the plastic insulation that protects electric cables, the **myelin sheath** protects the axon, so that the information does not leak. It is made of a material called myelin, which is made by the **Schwann cells**. The glial cells wrap themselves around the axon, making an extra cover. We call the gaps between Schwann cells the **Nodes of Ranvier**. These allow the neurone to recharge itself.

Synapses

A synapse is where one neurone meets another to pass on the information. A neurone will have synapses with dozens of others. They also act like switches, slowing down or speeding up news travelling through the nerves. The brain makes new synapses between different neurones all the time, and that is how we make new memories and learn new things.

Action Potential

This is a tiny electric current that passes along the wall of the axon. When there is no message to be passed along the nerve, there is sodium (Na⁺) outside the neurone and potassium (K⁺) and chloride (Cl⁻) inside it. This is the resting potential. When there is a message to be passed, Na⁺ rushes in and K⁺ rushes out, creating the action potential. After the message has passed, the two slowly switch places again, bringing the neurone back to normal.

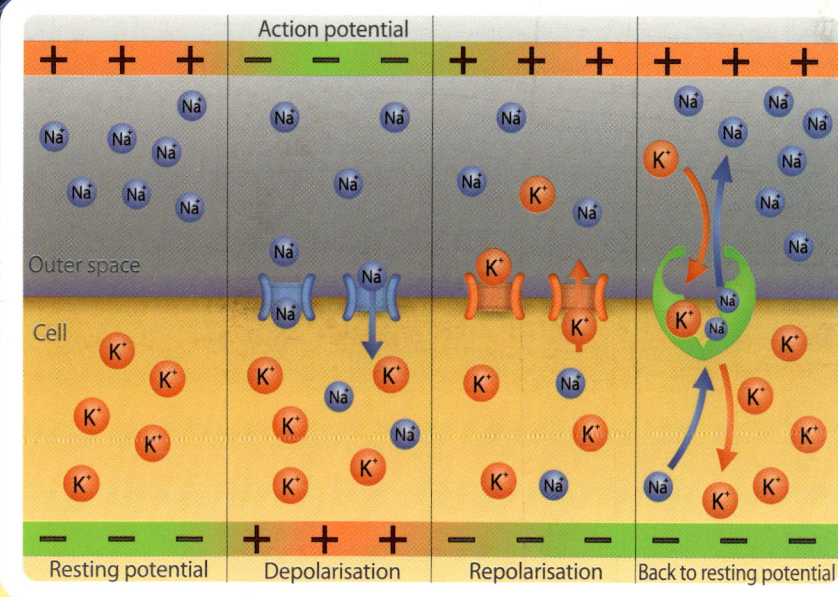

▲ Neurones take messages to and from the brain through action potentials

Mirror Neurones

These are neurones found in many parts of the brain. They help to bridge some of the senses with action. Scientists conducted experiments in which they found that these neurones become active when we see or hear something that is familiar to us. For example, these become active in ballet dancers' brains when they see other ballet dancers. These neurones help us learn things by seeing or hearing them, and also in expressing feelings like sympathy.

Incredible Individuals

Santiago Ramón y Cajal (1852–1934) was a Spanish neuroscientist who discovered that the neurone was the basis of the nerve cell. He developed techniques to 'stain' individual neurones and follow their path across the nerves. Many of the drawings he made are still used by medical teachers.

▲ Santiago Ramón y Cajal was the first Spaniard to win a scientific Nobel Prize

Your Body's Other Brain

The spinal cord takes messages from the organs of the body to the brain and back from it to the organs. It also takes messages to the muscles from the higher brain, but it does a lot more than acting like a courier. It also governs the sympathetic and parasympathetic nervous systems, which do not depend on the brain. It further handles automatic reactions of your body that the brain does not get told about, like your reflexes. That is why it acts as your body's second brain. Any damage to the spinal cord makes you unable to move some parts of the body, which doctors call paralysis.

Spinal Cord

The spinal cord runs entirely inside your backbone. Unlike the brain, the white matter is outside, and the grey matter is inside. The white matter is made of the axons and dendrites of sensory neurones coming in from the various organs and the skin, motor neurones going out to the organs and the skin, and neurones travelling through the length of the spinal cord, connecting to the brain stem. The grey matter is made of the cell bodies of these neurones, and small connecting neurones called **interneurones**.

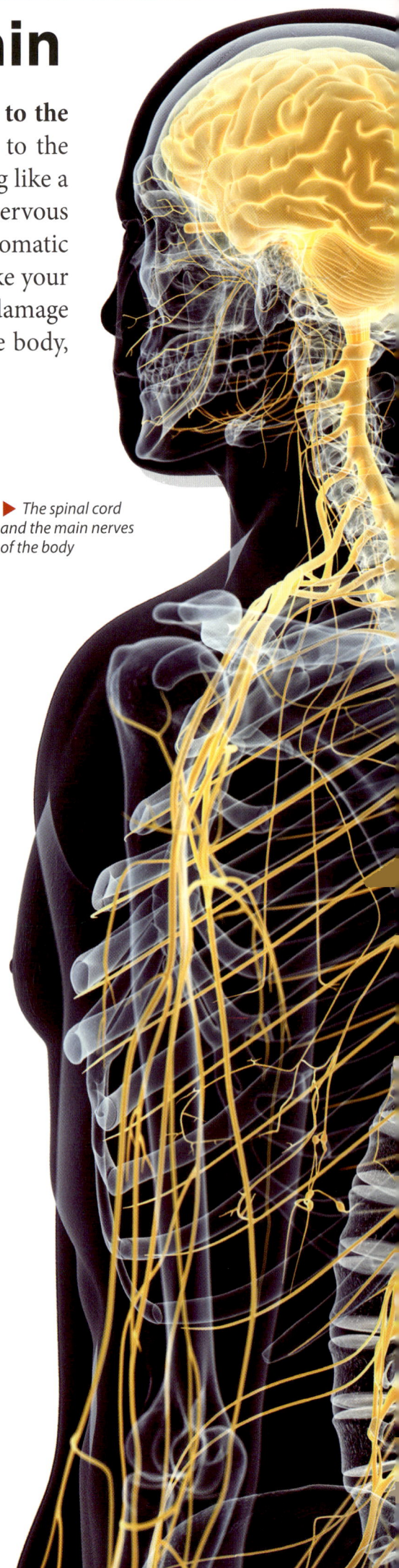

▶ *The spinal cord and the main nerves of the body*

In Real Life

Does the octopus have a backbone? No! That is precisely why it is called an invertebrate marine animal. This feature, along with them being extremely intelligent and curious creatures, makes octopi incredible escape artists. For instance, Sid, an octopus who was in captivity in New Zealand, escaped his tank multiple times before finally being released by the aquarium officials as they were fed up.

▲ *When bored, octopi are known to amuse themselves—from juggling hermit crabs to short circuiting aquarium lights*

HUMAN BODY | BRAIN & NERVOUS SYSTEM | 19

Reflexes

Reflexes are reactions of the body that do not require you to think. These are actions you take out of fear, hunger, pain, or anything else that requires you to act quickly. During a reflex, the sensory neurone connects with an interneurone through a synapse, which connects to a matching motor neurone. For example, some neurones from your fingers that can sense very hot things are connected to neurones in your forearm that pull the hand away. So, when you touch a pie right out of the oven, you feel the heat and pull your hand away, dropping the pie. Your eye sees the mess and your hands get to work cleaning it up. That is a second reflex! But it is the one that goes through the brain.

There are two kinds of reflexes. The somatic reflexes connect the sensory organs with skeletal (voluntary) muscles. These often act as part of the sympathetic nervous system, which helps you to deal with sudden situations. You can train yourself to stop these reflexes.

Visceral reflexes connect the sensory organs to internal (involuntary) muscles and you cannot control them. These are often part of the parasympathetic nervous system, such as feeling hungry after seeing a cake.

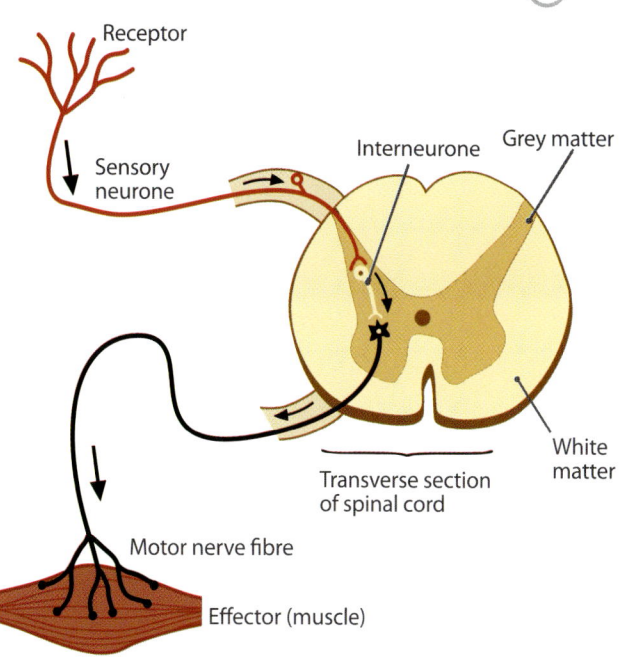

▲ Spinal reflexes allow you to act without waiting for the brain to decide

Keeping Your Two Brains Safe

The brain and spinal cord are not supplied with blood directly. Instead, the cerebrospinal fluid keeps them nourished. Blood capillaries in the brain are surrounded by tiny cells called glia. The glia makes up the blood-brain barrier and makes sure that nothing other than water and minerals can leave the capillaries and enter the cerebrospinal fluid. Proteins, large fats, and germs cannot cross this barrier. Glucose is moved into the brain by active transport.

Isn't It Amazing!

Chameleon tongues have some of the fastest reflexes in the world. Once it has seen its prey, a chameleon can shoot its tongue out at 2.6 kmps to catch it.

▲ The ventricles, full of cerebrospinal fluid, nourish the brain

◀ Chameleons have some of the fastest reflexes in the world

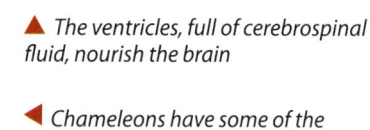

Switching Your Nerves Up and Down

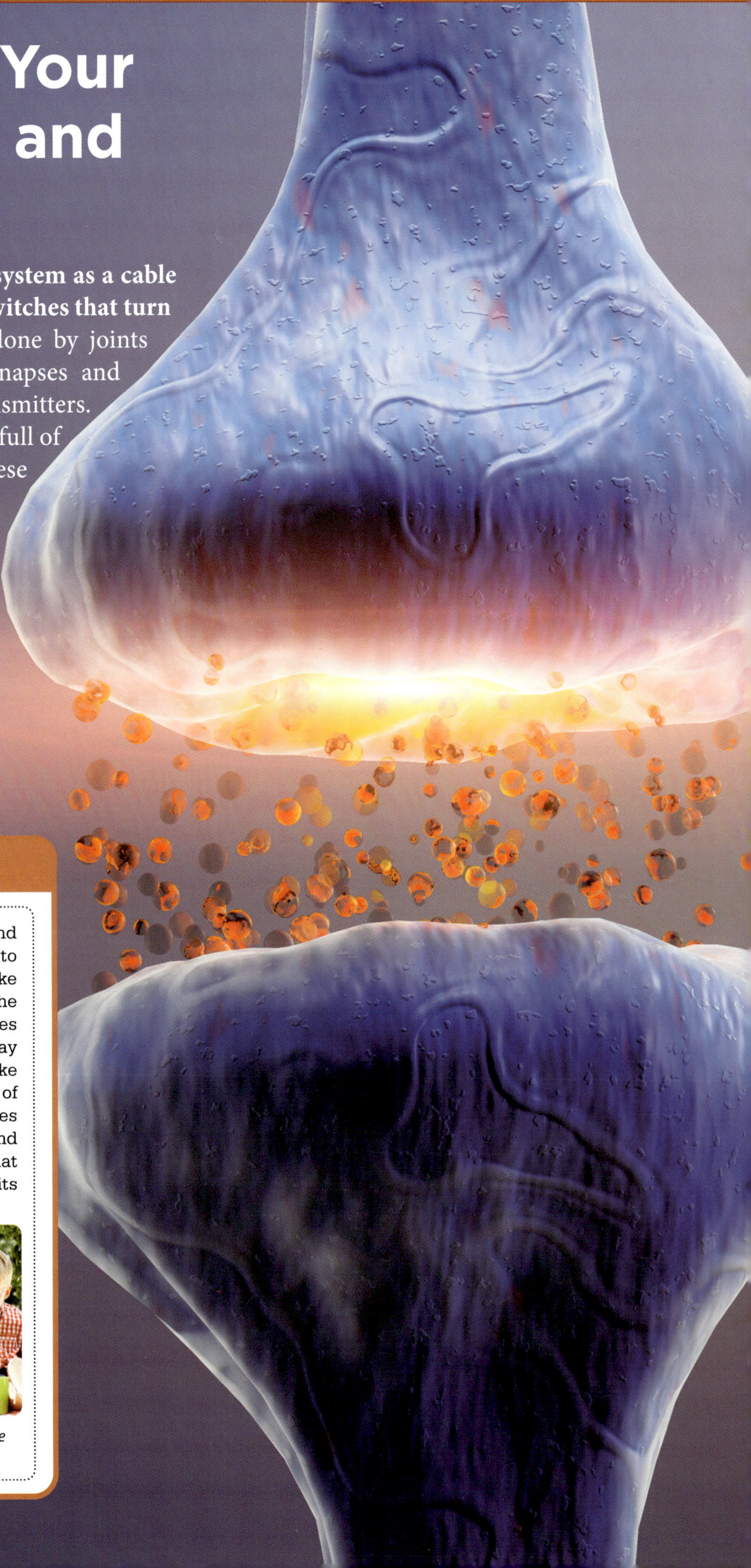

If you think of the nervous system as a cable network, it has got to have switches that turn things on and off. This is done by joints between neurones called synapses and chemicals called neurotransmitters. The brain and spinal cord are full of synapses. In the spinal cord, these are important for reflexes to happen smoothly. In the brain, they help it to make up new ideas, thoughts, and memories, but we do not know how that happens. However, it does seem like the more synapses that animals have, the smarter they are.

In Real Life

The brain stem, midbrain, and frontal cortex all work together to help you decide whether you like some things or not, through the reward pathway. The synapses of the neurones in this pathway use **dopamine**. If you really like something, like chocolate, a lot of dopamine is released. Sometimes this goes out of control and becomes an addiction. Drugs that block dopamine from tagging its receptors are used to treat it.

▲ Dopamine helps control what you like and what you do not like

Synapses

Synapses link the axons of one nerve with the dendrites of the one after it. When an action potential reaches the end of its axon, it triggers tiny sacs at the tip, called synaptic vesicles, that are filled with chemicals known as neurotransmitters. These go to the end of the axon and burst, spraying the chemical into the space of the synapse. The chemicals land on the surface of the next neurone, where there are proteins called receptors lined up on the membrane. When the neurotransmitter touches the receptor, an action potential starts in the second neurone. Our brain may have nearly three hundred trillion synapses, 300,000,000,000,000 of them!

Synapses form when two nerves connect with each other. If they keep 'communicating' to each other, the synapse becomes a strong synapse, else it remains a weak synapse. If the nerves have not communicated in a long time, the synapse may be broken up. Some scientists believe that the formation of strong synapses helps in keeping memories and that the breakage of weak ones makes us forget things.

 # Neurotransmitters

These are the chemicals that help neurones 'talk' to each other. There are three key types of neurotransmitters based on their functions. Excitatory neurotransmitters turn up the action potential in the next neurone; inhibitory neurotransmitters turn down the action potential in the next neurone; and modulatory neurotransmitters send messages to many neurones at the same time and can also communicate with other neurotransmitters. Many neurotransmitters also act like hormones.

 ## Incredible Individuals

Camillo Golgi (1843–1926) was an Italian neuro-scientist and teacher of Ramón y Cajal. He discovered many features of the nervous system, including the 'Golgi Vesicle'. This is a tiny bunch of membranes within each cell, where things that need to be sent out of the cell (like neurotransmitters) are packaged into little bags called 'vesicles'.

▲ *For their work, Golgi and Cajal were jointly awarded the Nobel Prize in Physiology or Medicine in 1906*

Neurotransmitter	Location	Nature	Response
Acetylcholine	Heart	Inhibitory	Makes the heart slow down
	Skeletal muscle	Excitatory	Makes the muscle contract
Glycine	Spinal cord and brain stem	Inhibitory	Quietens the next neurone
Glutamic acid	Brain and spinal cord	Excitatory	Turns up the next neurone
Dopamine	Some nuclei of the brain	Inhibitory	Quietens the next neurone
	Some nuclei of the brain	Excitatory	Turns up the next neurone
Noradrenaline	Skeletal muscle	Excitatory	Makes the muscle contract
	Heart	Excitatory	Increases heartbeat
Serotonin	Grey matter of brain	Inhibitory	Quietens the next neurone

 # Electroencephalography (EEG)

You know that little electric currents are how our nerves carry messages from the sense organs to the brain and back to the motor organs. These can be measured by neuroscientists by using a special instrument called an electroencephalogram or EEG machine. Diseases like epilepsy and motor neurone disease can be diagnosed using this machine. By attaching other machines, a process called Functional Magnetic Resonance Imaging (FMRI) is used to find out which part of the brain becomes 'electrically active' when you are thinking about something, solving a math problem, or reading.

From Thought to Action

Sometimes we do things after carefully thinking about them, like writing the answer to math problems. Often, we do not think at all, like when we need to catch a ball in a hurry. All these actions are governed by the motor nerves. They carry the brain's decision to the muscles, which do as they are told. The polio-causing virus, poliomyelitis virus (PMV), attacks the motor neurones and causes paralysis in its patients. Another disease called Amyotrophic lateral sclerosis (ALS) also affects the motor neurones.

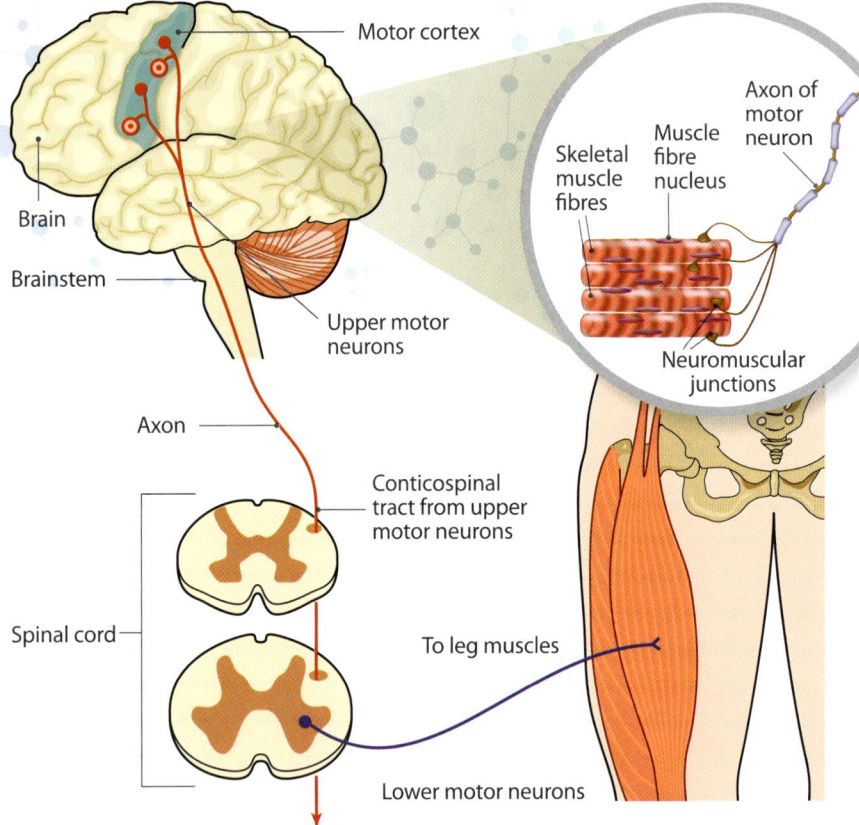

◀ The motor neurones finally tell the muscles what to do

Motor Neurones

The part of the brain that deals with actions is called the motor cortex. From there, upper motor neurones take messages to the spinal cord through the midbrain. **Lower motor neurones** start from the spinal cord and finally connect with muscles. Conscious actions need both, while reflex actions work only through the latter.

Muscles

Our muscles are made of lots of muscular fibre. Motor neurones make lots of synapses with them, that doctors call **neuromuscular junctions**. When a current comes through the neurone to the junction, it makes the muscle contract. In turn, the bone or organ that it is attached to is made to move.

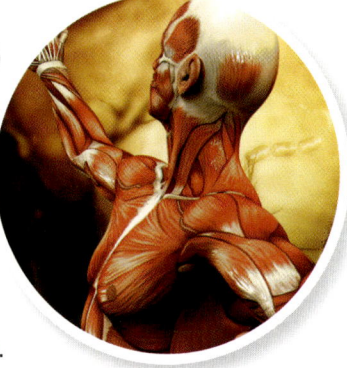

▲ Human muscles are made of lots of muscular fibres

In Real Life

An hour of play outside home helps train your nervous system to control your movements and speed up your reflexes. It also speeds up the ability to think on your feet. So, go ahead, drop that gaming console and pick up a football.

▲ Play is good for your motor nerves

Stress

Stress is how our nervous and muscular systems respond to changes in the world around us. Acute stress happens when there is immediate danger. The sympathetic nervous system wakes up the muscles, and prepares you to run away. Chronic stress happens when you cannot easily run away from a problem, like exams.

Making Memories

We still do not quite know exactly how our brain remembers things. Some scientists believe it has to do with the number of synapses in the brain. But we do know that memories come in two forms. The first are permanent ones, called engrams. You remember these for a long time, like your birthday; or a really terrible experience you might have had, like being pranked with a cake full of chilli. Others are called short-term memories, which you soon forget, like what you had for breakfast. You might forget it faster if you did not like what you ate.

Hippocampus

Did you know that **hippocampus** means seahorse in Greek? Well, whether it looks like a seahorse to you or not, it is very important in how you remember things. You have two of them, in each of the temporal lobes. They help you remember where things are—as in your spatial memory, mix many inputs into a single memory—so when you remember an event, you remember the sights, sounds, smells, and more; and also decide to remember some things for a long time such as a trip to Disneyland and forget other things, for instance, your dog having eaten your homework.

Incredible Individuals

London's taxi drivers have to learn the routes to nearly 25,000 streets to get a licence. This means the hippocampi in their brain grows bigger to help make room for all that memory.

▶ London taxi drivers have bigger hippocampi than regular people

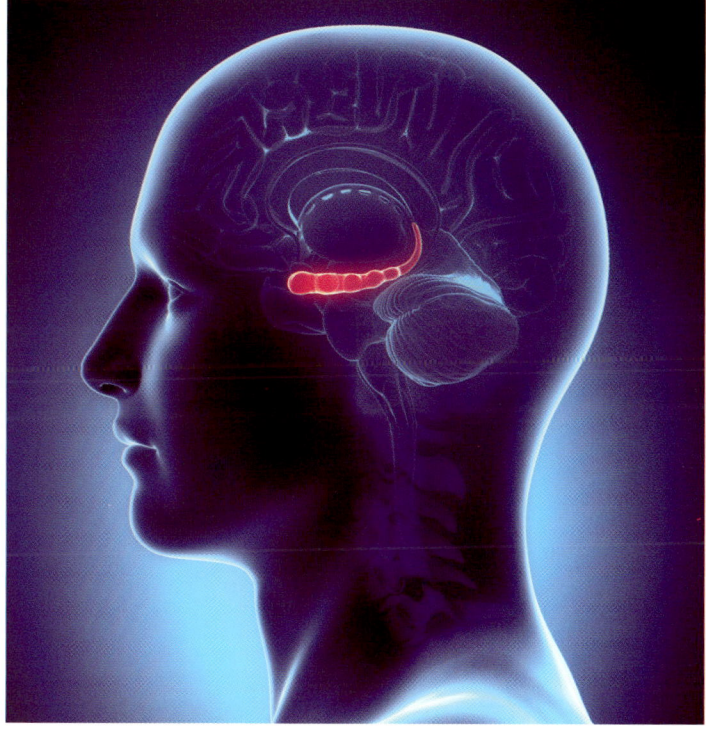

▲ Location of hippocampus

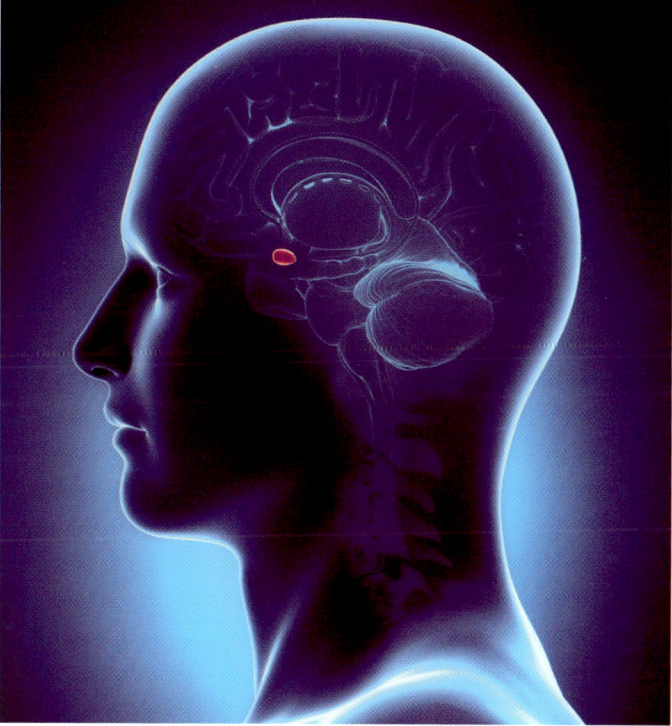

▲ Location of amygdala

Amygdala

This is a pair of tiny, almond-shaped parts of the temporal lobe, one on each side. The amygdala controls learning and memory along with the hippocampus. It also controls the reward pathway. This is the way in which the brain makes you feel good after eating something nice like, a pastry or doing something good such as helping someone in need. It is also involved in addictions to drugs and alcohol. It also makes you feel bad for things you didn't like to do, like eating broccoli!

Higher Thought
From Animal to Human

What happens when you see a delicious plate of food? How does the brain put all the senses together? How do we remember what the food tastes like, and how do we learn to like it? The answer to these questions is in the cortex. The cortex appeared only in animals such as birds and mammals. As human beings evolved, it became more complex.

Cortex

Humans have the largest cortex (*see diagram of human brain pp 6*) compared to body size. A human cortex has 16.3 billion neurones, compared to an elephant's cortex, which has only 5.6 billion neurones, even though it is twice as large.

Combining the Senses

Different parts of the cortex take care of different senses and motor actions. They connect to sense or motor organs via the inner brain and midbrain. The associated areas bring different senses together and that is how something becomes a multisensory experience. Broca's area and Wernicke's area are better developed in human beings than in any other living organisms. They help with language processing. Finally, the prefrontal lobe takes up more complex things like telling right from wrong, social behaviour, doing math, personality, and character.

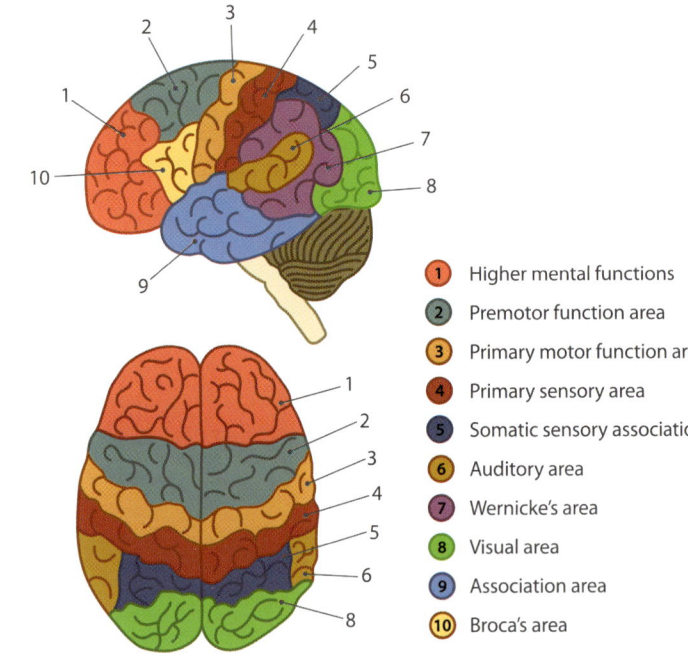

▲ *Different functional areas of the cortex based on its lobes*

1. Higher mental functions
2. Premotor function area
3. Primary motor function area
4. Primary sensory area
5. Somatic sensory association area
6. Auditory area
7. Wernicke's area
8. Visual area
9. Association area
10. Broca's area

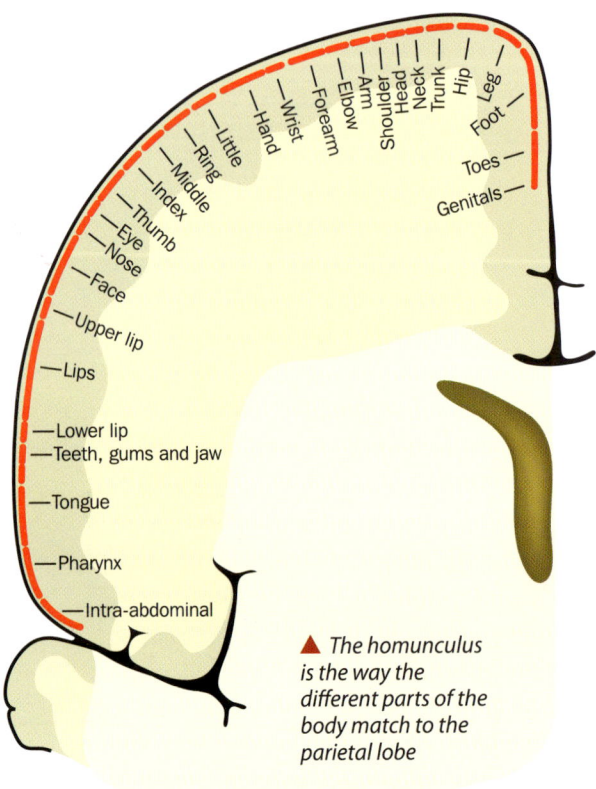

▲ *The homunculus is the way the different parts of the body match to the parietal lobe*

Homunculus

Homunculus is Latin for little fellow. It is a drawing made by neuroscientists that shows how much space the parietal lobe, which deals with touch gives to each part of the body. Some parts of the body, such as the lips and fingers are very sensitive to touch. Therefore, they have more nerve-endings and have a lot of matching space in the brain.

In Real Life

A stereotype is how we think of a group of persons or animals. For example, most people believe foxes are sly or cunning. Some stereotypes are useful, for example, if you see a lion, or a tiger, you better stay away. But many of them are harmful, and lead to negative thoughts like fear or bias against people of different backgrounds.

The Pituitary and Pineal Glands

When we are tired, we go to sleep. When we go out in the Sun, our body makes more **melanin**, giving us a tan. When we have been sufficiently rested, we wake up. How does our body know all this? This is where the endocrine system comes in. The organs of this system are called glands and make different **hormones**, which are chemicals that travel through blood and tell the other organs of your body what to do. The glands of the body, such as the liver, pancreas, thyroid, and adrenal gland are, in turn, governed by the pituitary gland. You can call it our body's chemical brain.

The Pituitary Gland

The **pituitary gland** makes hormones that control how other glands work. In turn, the pituitary is controlled by the hypothalamus, which contains centres for controlling hunger, thirst, body temperature, tiredness, etc. It makes the:

- ✔ Growth hormone, which makes your body grow.
- ✔ Prolactin, which helps a woman's breasts to make milk for her baby.
- ✔ Thyroid-stimulating hormones, which make the thyroid gland release the thyroid hormone.
- ✔ Adrenocorticotropic hormones, which make the adrenal gland release adrenaline.
- ✔ Antidiuretic hormones, which make the kidneys take up water filtered from the blood.
- ✔ Melanocyte-stimulating hormones, which make the melanocytes in the skin make melanin, which protects you from the UV rays of the sun.

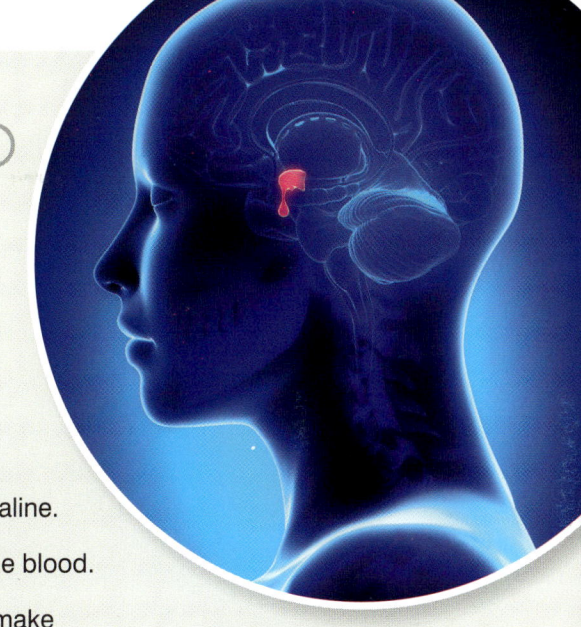

▲ *The pituitary gland is called the master gland of the endocrine system*

Pineal Gland

This is the brain's other gland, behind the hypothalamus. It makes a hormone called melatonin, which acts on the hypothalamus and tells it to go to sleep. You can also get melatonin from fish, eggs, nuts, and mushrooms. The pineal gland makes melatonin all through the day and releases it into the blood as daylight fades. When light falls on you, the brain and intestines make serotonin, which tell the brain to wake up. Over time, the cycle of sleep and waking becomes regular, and less dependent on light. This is called the circadian rhythm.

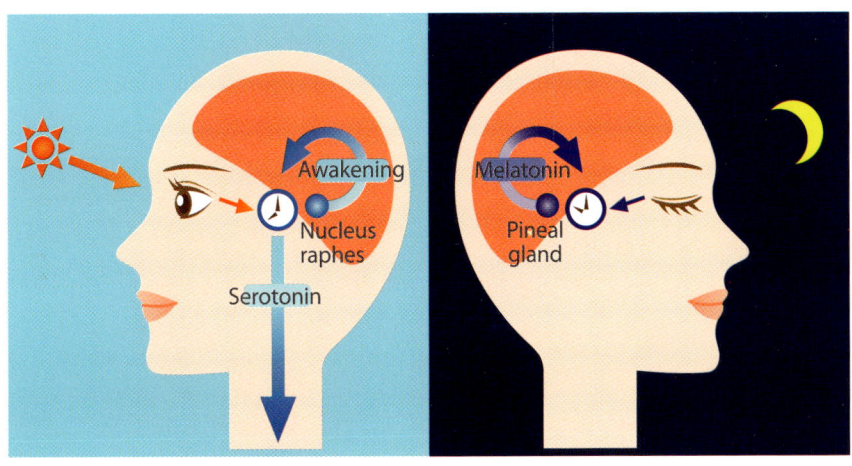

▲ *Melatonin makes you sleep; serotonin wakes you up*

💡 Isn't It Amazing!

In many birds, reptiles, and mammals, the pineal gland is very important. The pineal gland helps these animals to know the right season to mate and reproduce. Removing this gland makes them unable to breed in the right season. In some species, the pineal gland can also detect light and make melatonin accordingly. So it is sometimes called the 'third eye'.

Higher Thought II

The brain is an organ just like any other. It too can fall sick. But unlike other organs, illnesses of the brain manifest in three ways. The symptoms could be physical, such as uncontrollable movements or paralysis; mental, like loss of memory or the inability to pay attention; or emotional, such as excessive stress or sadness that does not go away.

The Brain and the Mind

The brain is the organ that governs your nervous system. The mind, on the other hand, is a complex set of faculties involved in perceiving, remembering, assessing, and deciding. While the brain exists as an organ with a concrete structure, the mind is invisible and thus, does not have a form like the brain. The mind is what the brain does. The study of the brain and the nervous system is called neuroscience. **Neurology** is the study of how the nerves work and what can go wrong with them. Psychology is the study of how your brain deals with the world, including how you think, your likes and dislikes, good and bad behaviour, and how you make decisions. Psychiatry is a field of medicine that studies what can go wrong with the brain, and how it can be set right. People who work in these fields are called mental health professionals.

▲ *Mental health problems can be treated by counselling, forms of therapy, and medication*

Depression

A depressed person tends to be sad for a long time. They think very poorly of themselves and feel hopeless, unhappy, and tired all the time. They find it hard to eat or sleep. It happens because the amygdala is not healthy and is not able to control serotonin and noradrenaline. It is treated with medicines called antidepressants.

Incredible Individuals

What is common between Winston Churchill, Beyoncé, Jon Bon Jovi, Johnny Cash, and Robin Williams? They have all suffered from depression in their lives. Remember, it is a mental health disorder and not a state of mind.

Bipolar Disorder

Someone with this disorder of the brain will feel depressed for a few days, and suddenly become very happy and excited on other days. This repeats again and again. It happens because of an underlying trouble in the release of neurotransmitters in the brain.

▲ Did you know that the music star Lady Gaga has bipolar disorder?

Anxiety Disorder

If some people worry about things too much, without any reason, they probably have anxiety disorder. Patients with this condition experience fear, sleeplessness, and mood swings. It happens because the thalamus and other parts of your brain that deal with stress, are not able to work properly. It is treated with counselling and drugs called anxiolytics.

ADHD

Know someone in school who has a hard time paying attention, can barely sit quietly, and gets bored easily? They might just have attention deficit hyperactivity disorder (ADHD). This happens because parts of the basal ganglia are unable to make enough dopamine and **noradrenaline**, which help pay attention and also act as 'brakes' on activity. ADHD cannot be cured completely, but counselling and some drugs help patients become calmer and more attentive.

In Real Life

It is important to seek clinical diagnosis of a mental health condition. However, an inability of the society to be understanding towards a diagnosed mental health patient, and discriminating or making fun of them, or associating shame with their condition causes social stigma. It is likely to make patients want to hide a mental health problem. Be nice to them and try to understand their problems.

▲ ADHD can go undiagnosed in children for years if their inability to focus on a task for long is misunderstood as a choice

An Attack of the Nerves
Epilepsy & Multiple Sclerosis

Some diseases affect the structure or working of the neurones, or both. These may not affect one's mental health, but they show up as physical symptoms like trembling and shivering or being unable to move. Some decades ago, such diseases were met with fear, scorn, or both. It was believed that such conditions were brought about by devils or the wrath of the heavens. Patients were subjected to electric shocks, lobotomies (*see pp 7*), and even tied in chains. However, modern-day medical science has evolved to help treat such conditions with medicine and surgery, if needed.

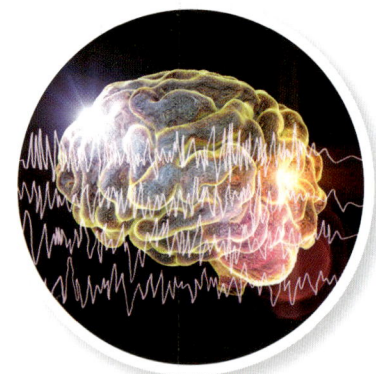

▲ *Around 1 in 5 of the world's children and adolescents have a mental disorder*

 ## Epilepsy

This illness affects the neurones of the brain. It may be in a small part of the brain or the whole brain. Both happen because the neurones suddenly lose control over their action potential, leading to currents running all over the white matter. During this time, the patient experiences muscle twitching, an inability to breathe, dizziness, and confusion, and sometimes fainting. Doctors call this a seizure. The patient may recover in a few minutes and may sometimes not remember what happened. Seizures may happen because of stress, sleeping poorly, or excessive consumption of alcohol, but unless they happen many times over, it is not considered epilepsy.

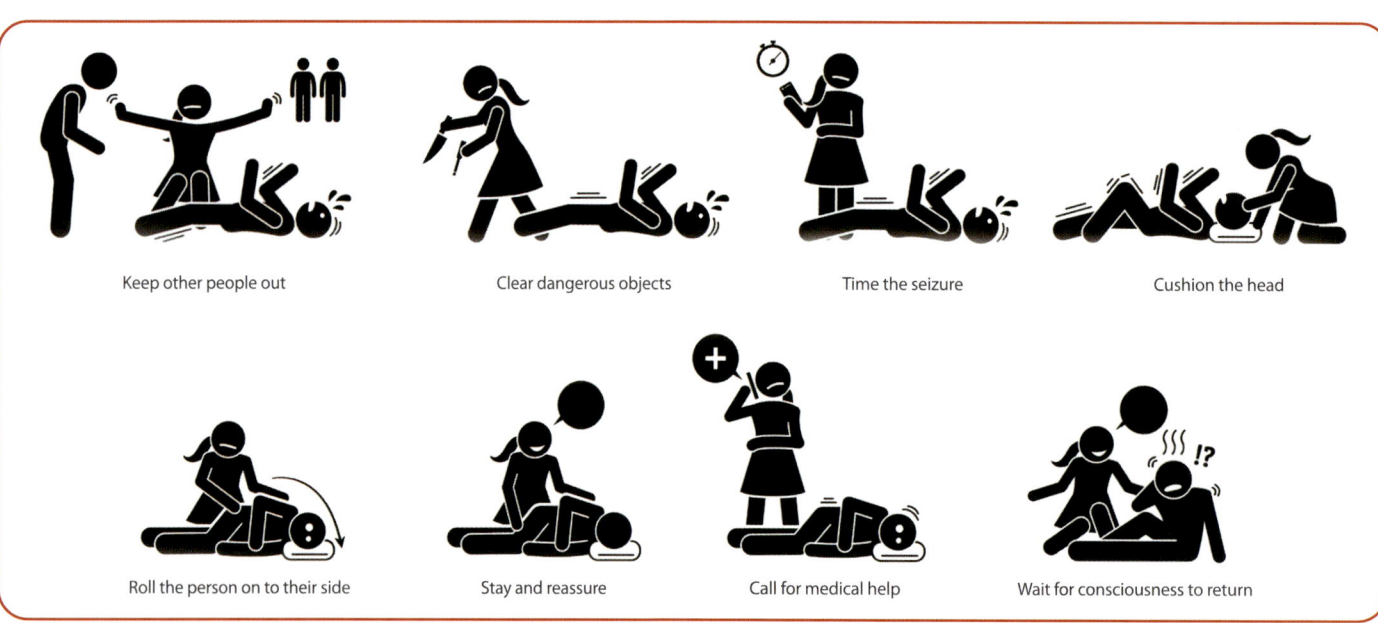

▲ *Here's what you should do when someone has a seizure*

 ## Multiple Sclerosis

This is a disease in which the myelin sheath of your neurones is lost over time. The neurones are not able to carry currents correctly. It causes you to feel dizzy and things to appear blurry. People can find it difficult to control their movements and can experience difficulty in remembering things. They slowly lose their ability to walk or use their hands.

In Real Life

It is believed that one should put something inside a patient's mouth when they are having a seizure, so that they do not swallow their tongue and choke themselves. However, doctors advise against this practice, so you should not put anything in their mouths.

Sleep Healthy and Wake Up Healthy

Your brain does not entirely sleep when you do. But it needs you to sleep, so that you can turn the day's experiences into memories, give your muscles and nerves some rest, and let your body grow. This rest is also important for your body to fight disease. Melatonin, made by the pineal gland, tells the brain when it is time to sleep at the end of a day. Over time, it becomes a habit, known as your circadian rhythm. Breaking the habit causes sleep deprivation. Stress or illness may make it difficult for your body to follow the rhythm.

Good Sleeping Habits

It is a good idea to start by maintaining a consistent sleep and wake-up time. Another point to bear in mind is to dedicate an hour before sleep to calming activities such as listening to soothing music or reading a book.

▶ Good sleep habits keep your body and brain healthy

▼ This girl has fallen into the habit of sleepwalking. Her motor neurones are active, but her sensory neurones are not!

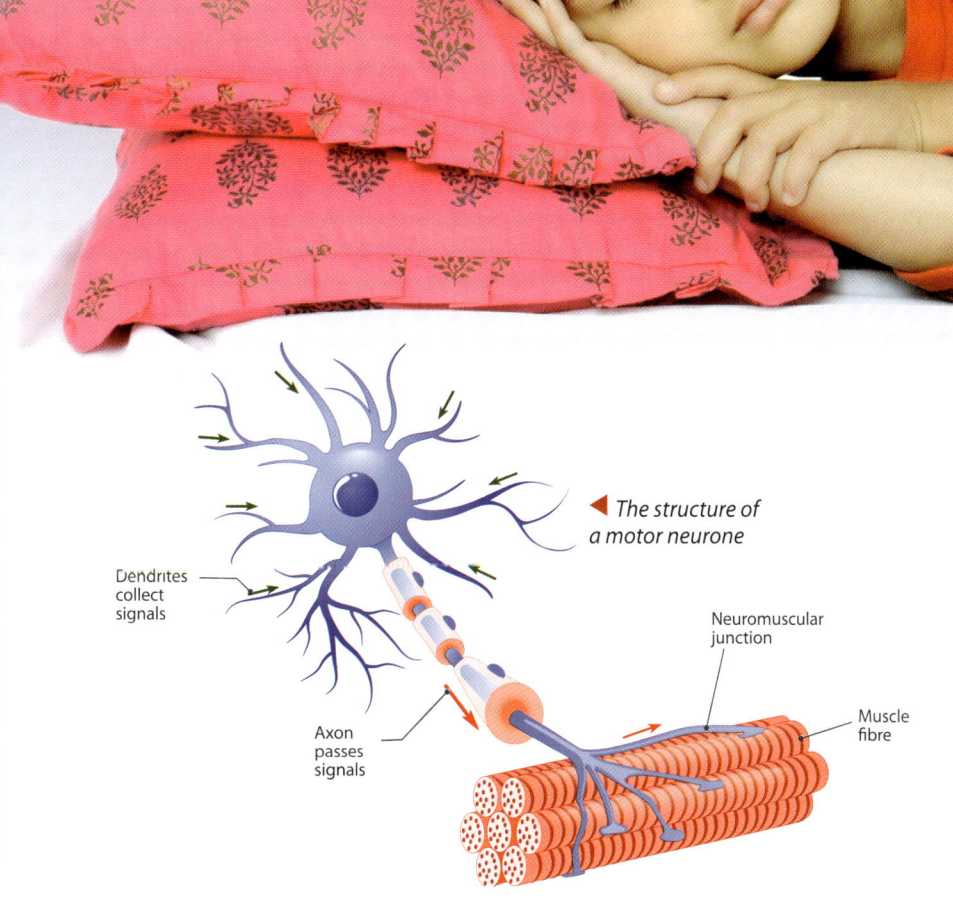

◀ The structure of a motor neurone

Dreaming and Sleepwalking

Did you know that when you are dreaming, your eyes dart about very quickly? This is called **rapid eye movement** or REM sleep. Though your senses are dormant, your higher brain centres are awake and experiencing things that are not happening in reality!

At other times, that is, in a **non-REM sleep**, your higher brain senses are also asleep. In some people though, especially children, the parts of the brain that control movement, that is, the motor cortex are awake. These children may get up from bed, start walking about, and even eat from the fridge! This is called sleepwalking, or somnambulism. It often goes away as you grow up.

A Healthy Mind in a Healthy Body

Though the brain occupies only 2 per cent of your body's weight, it burns upto 20 per cent of the calories you eat and drink. This is because your brain is always at work, even if you are sleeping, which is why you get dreams. But the brain and the rest of the nervous system can only use glucose as a source of energy. The hypothalamus makes sure that there is always enough glucose in the blood to keep the brain fed, even if other tissues have to do with less. If there is not enough glucose, the liver and muscles will turn the glycogen stored in them to glucose to keep the brain nourished.

The Nerves and Fatty Acids

Omega-3 and omega-6 fatty acids are very important to keep your neurones healthy, as they make up the myelin sheath. They are also needed in the **retina**. You get lots of them from seafood, nuts, soybean, and canola oil.

So, tell your parents to get you a big banana split with lots of nuts in it. The nuts give you vitamins, minerals, and fatty acids, the milk in the ice cream gives you calcium, and the bananas give you glucose and potassium. But remember to play and exercise a lot afterwards, so your motor nerves are in good health too!

◀ Foods rich in omega-3 and omega-6 fatty acids

Vitamins and Minerals

Your body needs a lot of vitamins and minerals to keep the nervous system working well. Iron and many vitamins are needed to make neurotransmitters. The minerals potassium, sodium, and chloride, which you get from salt, aid the normal working of action potentials. Calcium is needed for the synapses to release neurotransmitters properly.

▼ Foods rich in potassium are good for your nervous system

In Real Life

Ever called someone who acted crazy 'nuts' or 'gone bananas'? Something that is odd is called 'fishy'. But all these things are actually good for you, for they keep your nerves healthy and make you smarter.

▲ Bananas also support heart health

Improve Your Mental Health

Though we still do not have answers to all the illnesses of the nervous system, indulging in activities that help you unwind can be very fruitful for maintaining a good mental health. This could include anything: from the use of colours and paints to bring your ideas to life, to playing with your pup. Science also backs the use of varied forms of therapy to diagnose, as well as treat mental health conditions in children.

Play Therapy

Play forms a crucial part of how you make sense of your world, as a child. It is also a means for children to express their feelings. Psychologists often use play therapy to aid children in processing complicated emotions such as grief resulting from loss of a parent or caregiver. Play therapy is also great to help you learn desirable behaviours and unlearn the faulty ones.

Art Therapy

It is not always easy to express your complicated emotions. In order to do so, you can draw art. Psychologists sometimes suggest drawing, painting, sculpting, clay modelling, etc. for children with depression, anxiety disorders, and so on. It is also used with children diagnosed with ADHD as well as low IQ.

Animal-assisted Therapy

Animal-assisted therapists have specially trained animals, who are friendly and warm, so you enjoy playing with them, and feel less worried or frightened about things. It facilitates a child in communicating their complicated emotions better with the therapist. So, when you are feeling bad, give your dog or cat a big fluffy hug!

▲ *Psychologists watch children play and figure out signs related to their mental health*

▲ *Art therapy helps you explore your inner 'experience'*

💡 Isn't It Amazing!

Catching a flight can be stressful. Before the flight, you are probably hoping you make it in time past security check or long queues. Airports around the world now have new kind of employees—dogs and cats who help passengers calm down and enjoy their wait, known as Emotional Support Animals (ESAs).

◀ *Cats, dogs, horses, bunnies, and even goats help with animal-assisted therapy*

Word Check

Amygdala: It is the part of the brain that controls emotions and memory.

Atrophy: It is the shrinking of a tissue due to illness.

Axon: Part of a neurone that takes the action potential to the next neurone

Basal ganglia: It is the nuclei in the inner brain that connect with the higher brain and help with decisions.

Cone Cells: They are the cells in your eye that make out different colours.

Cortex: It is the part of the cerebrum that does higher functions such as calculations after making sense of stimuli.

Dendrites: The parts of a neurone that take the action potential from the previous neurone or sensory organ.

Dopamine: It is a neurotransmitter that helps regulate the movements a person makes as well as their emotional responses to a stimuli.

Ganglia: They are the clumps of neurone cell bodies in the brain and spinal cord.

Grey matter: It is the part of the brain and spinal cord that is made of the cell bodies of neurones.

Gustatory receptor cells: The cells in the tongue which pick up taste and relay it to the sensory nerves.

Gyri: They are the folds of the cerebral cortex full of grey matter.

Hippocampus: Part of the temporal lobe responsible for memory.

Homunculus: It represents the way the sense of touch from each part of the body is placed on the parietal lobe.

Hormones: They are the chemical messengers that travel through blood.

Interneurones: They are neurones in the spinal cord that control reflexes.

Involuntary movements: They are the movements within your body, such as heartbeats, which you cannot control.

Lower motor neurones: They are the neurones that carry messages from the spinal cord to the muscles.

Medulla oblongata: It is the part of the brain that connects it to the spinal cord.

Melanin: It is a natural skin pigment that gives our eyes, skin, and hair their colour.

Meninges: It forms the protective covering of the brain and the spinal cord.

Mirror neurones: They are the neurones that make the body do what it sees.

Myelin sheath: It is the protective covering around axons of neurones.

Neurology: It is a field of medicine that studies how the nerves work and what can go wrong with them

Neuromuscular junctions: They are the synapses between neurones and muscles.

Neurones: They are the cells that take messages from one part of the body to another.

Neurotransmitters: They are the chemicals in synapses that turn up or turn down neurones.

Nodes of ranvier: They are the gaps in the myelin sheath covering an axon.

Non-REM sleep: It is the kind of sleep when you are in deep sleep or sleepwalking.

Noradrenaline: It is a hormone and a neurotransmitter.

Nuclei: They are the clumps of grey matter in the brain and spinal cord.

Olfactory bulb: It is the part of the brain that deals with smell.

Olfactory receptor cells: They are the cells in your nose that catch smelly chemicals.

Papillae: They are the taste organs in your tongue.

Photoreceptor cells: They are the cells in the retina that are sensitive to light, made up of rod cells and cone cells.

Pituitary gland: It is the gland that makes many hormones, it is controlled by the hypothalamus.

Rapid eye movement sleep: It is the part of sleep when you are dreaming.

Retina: It is a thin membrane which lines the back wall of the eye.

Rod cells: They are the cells in the retina that sense whether it is light or dark.

Schwann cells: They are the cells that make myelin and cover the axons of neurones.

Sulci: They are the grooves between folds (gyri) of the cerebrum.

Synapses: They are the connections between neurones that act as switches.

Synaptic vesicles: They are the bags at the end of axons filled with neurotransmitters.

Ventricle: It refers to a chamber of the organ, in this case, the brain.

White matter: It is the part of the brain and spinal cord made of the axons and dendrites of neurones.